EMERGENCY SURVIVAL PACKS

How to Prepare 72-Hour and 14-Day Family Evacuation Kits

I0807926

EMERGENCY SURVIVAL PACKS

How to Prepare 72-Hour and 14-Day Family Evacuation Kits

Blair D. Jaynes

Copyright © 1982, 2004 By
HORIZON PUBLISHERS & DISTRIBUTORS, INC.

All rights reserved.
Reproduction in whole or any parts thereof in any form or by any media without written permission is prohibited.

Fourth Printing: July 2004

International Standard Book Number: 0-88290-202-4
Library of Congress Number: 82-81293

Horizon Publishers' Catalog and Order Number: 1230

Printed & distributed in the United States of America by

CEDAR FORT

Mailing Address:
925 North Main Street
Springville, Utah 84663

Local Phone: (801) 489-4084
Toll Free: 1 (800) SKYBOOK
FAX: (800)489-1097

E-mail: skybook@cedarfort.com
www.cedarfort.com

About the Author

Blair Dee Jaynes, age 39, happily married to Karren B. Archibald of Edmonton, Alberta, Canada, and the father of four lovely children (Jonathan 13, Christopher 12, Suzanne 10, Joshua 7), is the first-born of the ten children of Robert Dee Jaynes and Violet F. Peterson, who now reside in Salt Lake City, Utah.

Born in Sacramento, California, Blair graduated from Idaho Falls High School in 1961. After a three-year tour in the U.S. Army in Germany and two years on a church mission in England, Blair entered the University of Utah, obtaining his B.S. degree in Psychology with honors. At that time, J. Reuben Clark Law School at Brigham Young University was just forming its charter class and Blair joined their ranks, graduating in 1976.

After another two-year tour with the U.S. Army as an attorney, Blair entered the world of corporate law, practicing for three years with the J. R. Simplot Company in Boise, Idaho. He is currently a Captain in the U.S. Air Force Judge Advocate General's Corps, serving as the full-time Staff Judge Advocate for the Idaho National Guard.

Blair's military training and experiences uniquely qualify him in the field of survival living. Having been trained in airborne and special forces operations, as well as combat engineering, Blair brings over twenty years of experience to the subject of survival in hostile environments. He was awarded the Humanitarian Service Medal for services performed during the Teton Dam disaster in 1976.

Contents

Preface

As the eldest of ten, I took it upon myself to assist my brothers and sisters, all now with families, with their temporal preparedness programs. In so doing, I discovered that there was not in existence a complete and definitive list which one could readily turn to for help in planning a comprehensive and portable short-term survival supply of both food and non-food items—items which one would require to survive a sudden calamity at some level above mere existence.

There are many good works on food "storage" and survival "techniques," but nothing I could find which would address the need to plan and organize what I call "family emergency survival packs"—portable packs which would contain sufficient food and essential non-food items to sustain a family through the first two weeks of a catastrophe.

Drawing heavily on my training and experiences in the military, as well as Scouting and camping lessons and experiences, I set out to develop what is contained in this small work. Friends and family alike have been of great assistance and support as I fine-tuned the lists. My sincere appreciation and gratitude goes to them. Of course, nothing in life is final; this work will, undoubtedly, undergo further refinement. I would hope that the reader will accept this effort for what it is—a labor of love, given to all who wish to accept it.

The Author

SHOCK-PERIOD SURVIVAL

The family emergency survival packs described in this pamphlet are designed to contain all of those items which are essential to the survival of you and your family during an initial 14-day "shock" period following a major catastrophe, whether it is a natural disaster, nuclear attack, or civil disorder.

It is assumed that you and your family survive the initial disaster which has now disrupted your normally ordered and interdependent society. You and your family must now survive the chaos and lack of services which are often the aftermath of disaster. In many cases the disaster itself will require you to flee the comfort and security of your home. Possibly the turmoil or lack of services which follow will force you to remove your family to a safer or more comfortable location.

One only needs to recall the pictures of refugee hordes fleeing their homes in time of war or after an earthquake to realize the immediate limitations which will be placed upon transportation and the delivery of goods and services. Therefore, the emergency supplies set out in the lists that follow are designed to be carried by your family. One can readily see the limitations inherent in a 50-pound sack of wheat under those circumstances. While these lists contain items which may not be used in every case, they should all be kept in your family survival packs, as the probability of their being needed in an emergency is high, and they would be critical to your survival if you did not have them.

The lists are organized so that you can plan your emergency supplies, regardless of your family size or the ages of your children. A single person (a family of one adult) can make good use of the lists also. All items from the Basic Family Unit List will be required for every family. To these items are added one set of the items from each of the other lists (Adult, Children, and Babies) for each family member in that age category. For example, a married couple with two children, ages 5 and 2, will prepare their family emergency survival packs as follows: 1 complete set of items found on the Basic Family Unit List; 2 complete sets of items found on the Basic Adult-Member list; nothing from the Basic List for Children (ages 8 to 11); 1 complete set of items from the Basic List for Children (ages 4 to 7); 1 complete set of items from the Basic List for Babies (ages 0 to 3).

You will note that the lists are designed to reduce to a minimum those items which cannot be shared by family members. There is little or no redundancy in the lists; sharing becomes a necessity. This is to reduce the weight of the packs to the minimum required for survival. Because of this, it is important that nothing is left out of your packs that is listed as required for the members of your family.

In developing these lists I have assumed that you will only have a short period of time to prepare prior to being forced to leave the comfort of your home. Therefore, the packs are designed to contain everything you and your family will need, even clothes. Of course, if you have time, you should put on clothing that is appropriate for the climate from your closets, and also provide yourselves with an extra cushion in your clothing supply. Remember, it is better to dress too warmly than to freeze later; you can always take off outer layers if necessary. Also, if time permits, you could add food and other survival items to your packs prior to leaving, but don't count on it.

The purpose of the family emergency survival packs is to have, in one location, all of those essential items your family will require to survive during the initial two-week period following a disaster. The nature of most disasters, for which we should prepare, will not permit us the luxury of going around the house gathering up those items, so do *not* expect you will have that chance. In fact, you may be forced to flee your home without the opportunity to retrieve your packs. In order to minimize this danger, you should keep your packs fully stocked and stored in a safe, secure location where they can be reached quickly; preferably near a window on the ground floor. You could then return and recover them easily.

Portable Packs

Start immediately to obtain the items on the lists. Get copies of good basic survival literature (see the appendix for a suggested list of such books) and review them regularly. Become familiar with the survival advice contained in them.

If you do not have backpacks or duffel bags with shoulder straps to store and carry the survival items in, use pillow cases, or make bags or packs from heavy cloth or canvas. Even a large plastic garbage can to collect your items in is better, at least temporarily, than just leaving the items on the shelf.

Do not become discouraged. You will be surprised how quickly and cheaply you can obtain most of the items on these lists. Check items off the lists as you obtain them. Set a family goal to acquire a few high-priority items every payday or whenever you get a few extra dollars. Have a garage sale or some other family fund-raising project to raise the extra money you need to complete your packs. If you truly want to prepare your family by completing your family survival packs, you will find a way to do it. Perhaps nothing could be more important for your family's future temporal and spiritual salvation. Yes, spiritual salvation, for how can you or your family continue to be God-fearing Christian saints when you are burdened by the temporal hardships, starvation, and even death, of your loved ones. Sure, it can be done—Job did it, but how much better it would be if you could provide temporal comfort, even on a survival level, to your loved ones because you were wise and prepared for the trials ahead. Because of your wisdom, planning and preparation, you and your family will have the strength to go forth and accomplish your mission, even in times of trouble.

You may want to form a family organization (if you don't already have one) with your brothers, sisters, cousins, etc., to help one another prepare your family packs. You will also reap the benefits of being able to purchase items (like freeze-dried food) in larger quantities. Shop around—check prices at several supply houses and discount stores that sell sporting goods and camping gear to get the best price. Watch for annual sales of camping gear. Make this a family-wide project with a high priority and you will be happy and secure with the result of your efforts.

As you build your packs, you must load them so that they can be carried by the members of your immediate family. They will not be light. You should not overburden younger members. Remember, they will be on survival rations, not steak and eggs. Fatigue can be debilitating, and lead to illness. Basic family-unit items, and items for those unable to carry their own pack, must be distributed to those family members carrying a pack. Your family's age, size, or other circumstances may require you to improvise some other way to transport your survival packs. An "A-frame" drag or, better yet, a handcart made out of bicycle wheels, could become part of your portable survival pack plan in that case. If your circumstances warrant such back-up devices, you should build them now, while you have the time and the resources.

You may be fortunate enough to be able to use your family vehicle in a crisis, if even for a limited time and distance. Again, don't count on such luxuries in your planning, but it would help to have a little extra fuel stored just in case you are able to use your car. At least you could try to keep your vehicle's fuel tank always half full.

List Testing

After you have collected the survival items your family requires, based upon the applicable lists, you should try the packs out. Test them. Take your family emergency survival packs into the hills and prove to yourself and your family that you can survive out of them. After this test you should evaluate the packs. Were you able to transport *all* of the essential items? Were there any items in your packs which you did not, and felt that you would not, under *any* circumstance, need? Were there items which you did *not* have in your packs which you feel would be absolutely essential to your family's survival?—not just "nice-to-have" items, but truly essential materials?

Modification of the lists after such a test and evaluation should certainly be done to fit your family's particular circumstances. I seriously doubt that you will delete any item already on my lists, however, modification, addition or deletion may very well be warranted by your circumstances. If so, go ahead and make those changes, being careful to consider all of the consequences of such decisions. No two families are alike. What is "essential" to one may be excess baggage to another. While these lists have been carefully developed after much thought, experience and research, they are not foolproof. They should be used as a guide, with careful independent thought and consideration of your own family's situation.

Remember, whatever you do, that the ultimate goal in putting your own family survival packs together is SURVIVAL. Do not add any "nice-to-have" items. They will only place unneeded burdens upon your family and, worse still, you may be tempted to delete a truly "essential" item to make room for that unnecessary "goody."

Survival Sense

A word about "survival." Once you have completely prepared your family for "shock-period" survival in the aftermath of a disaster by organizing your emergency survival packs, do not expect that your survival is assured. You cannot just take your packs out into the woods, set up a tent, build a lean-to or roll out your sleeping bag, sit down and wait for civilization to return to normal in 14 days. Survival living is *not* camping. Your packs are only intended to give you a headstart. They are a base from which to operate and build. They *cannot* guarantee survival. You must be willing to work, and work *hard,* to see that your family will survive whatever disaster has befallen you.

You will note, for example, that you cannot store or carry enough water in the canteens on the lists to last you 14 days. In fact, the water in them will usually last just one day! However, do not try to ration water; permit family members to drink their water at their own pace. Obviously, then, if you are using your "survival sense" you will realize that you must quickly find a good water supply, and purify it, if necessary. Remember, also, that any dehydrated food you pack will be rather hard to digest dry! Wood for heating and cooking must be gathered. Shelters or tents must be erected. Sanitary latrines and garbage pits must be dug—cleanliness is mandatory to avoid disease. The disaster itself, whether hurricane, earthquake or nuclear war, will also present its own particular challenges to your survival.

Your entire effort must center around survival; you must eat, drink, think, sleep and dream SURVIVAL! There are no assurances that things will be ok in 14 days. You must start immediately to supplement your precious shock-period food supply with available resources in the area. Snare, trap, net, fish, gather edible plants and scrounge wherever and whatever you can to assist your family in its new, and hopefully temporary, goal of survival.

A final word about survival sense. There is no way that you can anticipate exactly what type of disaster may require you to use your emergency survival packs and force you to change your life style from the "TV dinner" to the "survival" mode. The real key to your survival is, ultimately and in one word, your *flexibility.* Be prepared to *not* be completely prepared. Don't panic just because things aren't happening as you thought they would. A nuclear war will require you to take certain precautions and adopt survival behaviors that an earthquake would not. A famine that triggers general civil disorder and riots will place different challenges ahead of you than a flood. Again, *do not panic!* If you keep your wits about you, use your native intelligence, God-given talents and acquired skills, along with lots of prayer, and have prepared beforehand as best you can, you will be able to survive the chaos that will follow any catastrophe.

Supplemental Scrounging

About scrounging. What is scrounging? Have you ever had to repair some broken household gizmo and needed a small thingamajig to hold the whatzis in place? Did you dig through that box (or 3) of odd bits and pieces of whatchamacalits and found one that, with a little bending and shaving, would do the trick, at least temporarily? If you have, then you know what scrounging is and how to do it. You can see a problem, analyze it for possible solutions, evaluate on-hand resources for potential solution material, and adapt that material to fit the required solution and solve the problem.

In the military we call it "field expediency"—getting a job done with what is available. After the holocaust you will not be able to get it from the local parts house; your ability to improvise from what is available will be your most valuable asset as you try to supplement your emergency supply with other needed items as time goes on. You should try to cultivate this ability. Learn to examine all available resources before you call the repairman. Scrounging in a survival mode is gathering those things available in your environment which you sense will be of some value to your family as they work at surviving.

But, be aware as you scrounge. You will not be the only one out there in the cold, cruel world who will be trying to survive. Competition for valuable assets will be great. While you should always keep to those standards of decency and honesty honored by all civilized people, there are those who will believe that the destruction (even temporarily) of our normal way of life will also signal the end of all law and order.

These people will feel that the current situation entitles them to become a law unto themselves. They will adhere only to the law of the jungle: "The survival of the fittest." They will use this "law" to justify thievery, robbery, assault, and even murder in order to get what they think they need in order to survive. Be prepared for this. Stick together with family and friends whom you can trust and who you know will hold to the Christian ideals you cherish. Avoid large groups of strangers and public camps. Isolate your family camp; on high ground, if possible. Halt any strangers who approach your camp before they get too close. Find out who they are and what they want; ask discerning questions which will disclose to you their real motives and intentions. If you have cause to be wary or unsure of their character or desires, send them packing!

Food Facts

A comment about the food listed for these survival packs is necessary. I have purposely not set out detailed lists of the food you should pack. This is an area which is full of opportunity for flexibility in both planning and execution. Just keep in mind the purpose of the packs and use what types of food are available to you. The list simply sets out recommendations. Suppliers of the type of food you choose for your family survival packs will have ample information on the quantities required for your particular family for a two-week supply. However, I felt that I should provide some ideas and guidelines.

Remember that these packs are *not* intended to substitute for or supplant any regular home food *storage* program you may have in your family. Certainly, you should try to store, in your home, sufficient food and water (in addition to what you have in your emergency survival packs) to meet your family's basic food requirements for a 1- or 2-year period. In ideal conditions you should be able to remain in the shelter of your home and utilize that basic food storage in times of crisis. If not, you will be prepared to meet the challenge of survival away from home with your survival packs. Perhaps your situation will permit you to return and use part of that basic food storage. Even if you can stay at home, the items in your packs will prove invaluable to your family's survival in times of crisis.

If you cannot obtain the recommended light-weight freeze-dried or dehydrated foods for your packs, or until you are able to acquire them, I suggest that you prepare Zip-loc or twist-tie plastic bags or plastic bottles containing the "basic four" food items from your regular storage. Wheat, honey, dry powdered milk and salt could also be supplemented with some dry beans, rice or other light, long-life foods you have available.

All food items should be distributed as evenly as possible throughout all of your family packs. Make sure that each pack contains some of each of the basic food items from your supply, whether "basic four" or dehydrated. You don't want to wind up with only one type of food, or missing one basic food group, if you lose a pack or must leave some packs behind. Also, a family member cut off from the rest of the group could survive for a time on his limited, but balanced, supply.

As you plan your food supply, do not plan on taking enough food, whether freeze-dried, dehydrated or basic-four type, to give all family members three square meals a day. You will not be able to carry a 14-day supply of food for your family, or even for yourself if you are single, if you plan three gourmet meals per day. Plan on taking only what your family can carry (or transport, if you have built a handcart), along with those essential non-food items on the lists. Start your supply with one nutritious meal, with moderate (not large) servings per day per person. If your family can carry more, add another light meal (with small servings), either breakfast or lunch, for each person for the 14 days.

There are benefits to be obtained, if your family can carry the heavier basic-four type foods, from mixing your dehydrated survival foods with food from your basic food storage. Your demand for water may be much less and you will be able to make the transition to the harder-to-digest wheat and beans menu more gradually. Of course, your family would benefit from the nutritional advantages of whole-grain wheat.

A sample menu from such a mixed supply of food could include a breakfast of whole-grain or cracked wheat cereal, with honey and milk made from powder. Dried fruit, beef jerky or nuts could be used at midday to bridge the gap between the light morning meal and the evening meal. Such quick-energy, no-fix foods should be placed in every pack. They will prove very useful to family members, especially children, during the first few hours, or days, prior to establishment of a camp or permanent water supply when regular meals can be cooked. Children should always have some of these items in their pockets or packs in case they become isolated from the rest of the family for a short time.

The evening meal could be a freeze-dried or dehydrated main course, such as beef stew, and a freeze-dried or dehydrated vegetable with water to drink. Or, dry beans could be used to make bean soup or chili. Rice could be used for variety either in the evening or for breakfast.

Remember, *remain flexible!* Use your imagination in planning your menu and food list. Keep it simple and lightweight. Remember that you will be supplementing this emergency food supply with local scrounging.

72-Hour Kits

A 72-Hour Kit is a lighter, less-comprehensive version of a 14-Day Emergency Survival Pack, designed for a much shorter emergency period. In the lists which follow, you may wish to use the left margin to identify the items you would include in 72-Hour Kits. The contents of such kits will vary based on the climate, time of year, environment (urban or rural), and intended evacuation sites of those making the lists. Many items, however, are essential to both 72-hour and 14-day preparation levels, so the basic lists remain the same.

A Final Word

The lists that follow are only a guide. They will help you plan for that emergency that we all hope and pray will never come. If you follow the advice set out here and in the lists that follow, you will be able to avoid the panic and confusion that beset most people in times of crisis. You will be confident; you will be prepared.

Certainly, you will be troubled, dismayed, and most probably afraid. That is only natural. But, you and your family *will* survive if you have prepared yourselves physically, mentally, and spiritually to survive. Make that your firm resolve. Use this small work as your guide. Go forth and prepare to meet tomorrow, whatever it may bring.

Basic Family-Unit List

All items on this list must be packed in the quantity shown, regardless of family size. Distribute throughout all family members' packs along with individual items from other applicable lists.

Number On Hand	Number Needed	Item
______	1	Silva liquid-filled compass
______	1	Small road map of U.S. & local area
______	1	Water purification kit (iodine or halazone) (Sufficient to purify cloudy water—approximately 8 tablets/gallon for a total of 7 gallons for each adult and 3 gallons for each child age 4 to 11.)
______	1	Snake bite kit
______	1	First aid kit (family size with the following items *added,* if not already in the kit: large compress bandages, adhesive tape, scissors, band-aids, antibiotic ointment, aspirin, triangle bandages, thermometer, paragoric, Pepto-Bismol or Kaopectate, hand lotion, vasaline, insect repellent, sun-screen lotion, foot powder, pediatric cough syrup, cold tablets or capsules, vitamin C (ascorbic acid), consecrated olive oil, and any prescription medication regularly required by family members.)
______	1	Fire starter kit
______	1	Small hand axe
______	1	Small bow saw
______	1	Folding camp shovel
______	1	Aluminum family cook kit (with the following items *added,* if not already in the cook kit: 1 fry pan, 1 large pan, 2 small pans, lids, handles, enough plates, cups and/or bowls for each child ages 0 to 7 in the family, pancake turner, plastic 2-quart pitcher w/lid, can & bottle opener, large spoon & barbeque fork, and an old pair of hot pads or mits.)

Basic Family-Unit List (continued)

Number On Hand	Number Needed	Item
______	1	Small plastic bottle of liquid dish soap
______	1	Heavy-duty scrubbing pad (square 3-M-type)
______	2	Cotton dish towels
______	1	Small dish rag or sponge
______	1	Small roll heavy-duty aluminum foil
______	1	Small package large Zip-loc bags
______	6	Large heavy-duty garbage can-size trash bags
______	1	Collapsible canvas or plastic water pail (2-5 gal.)
______	1	Set of scriptures (paperback or old)
______	1	Unused journal for recording your experiences
______	1	Hymn book (pocket sized or used)
______	1	Copy of your family's 4-generation genealogy sheets
______	4	Pencils
______	2	Inexpensive ball-point pens
______	100 ft.	¼-inch nylon or rayon rope
______	1	.22 caliber rimfire semi-auto or bolt action repeating rifle
______	200	Rounds of .22 long rifle hollow-point ammunition
______	1	Small break-down fishing rod & reel
______	1	Small fishing kit with several sizes of hooks, weights, lures, flies and floaters
______	6	White camp candles
______	1	Large (but light) battery lantern (w/spare bulb & battery) *or* small propane lamp
______	1	Small battery-powered radio (preferably with short-wave) and 1 set of spare batteries
______	1	Clean white flat twin-size sheet (for bandages, rags, and towels)
______	4	Pair long heavy-duty boot laces
______	1	Dental floss dispenser with floss
______	1	Small hair brush
______	1	Copy of the Boy Scout Field Book
______	1	Set of basic survival literature (see Appendix)
______	1	Folded 3 × 5-foot American flag

Basic Family-Unit List (continued)

Plus food: Pack sufficient freeze-dried, dehydrated, or basic-four storage foods to feed the entire family unit for 14 days. No cans or glass bottles—plastic bags and containers only.

Use this space to plan and list your food requirements:

Number On Hand	Number Needed	Item
______	______	______________________________
______	______	______________________________
______	______	______________________________
______	______	______________________________
______	______	______________________________
______	______	______________________________
______	______	______________________________
______	______	______________________________
______	______	______________________________
______	______	______________________________
______	______	______________________________
______	______	______________________________
______	______	______________________________

Also pack enough multiple vitamin/mineral supplement (1 per day times the number in the family) and liquid vitamin drops for babies.

If you have room, and can handle the weight, you may now consider adding quick-energy and basic food items, such as: Bisquick-type flour mix, dried fruit, jerky, nuts, pre-sweetened Kool-Aid-type drink or candy.

______	1	*Optional:* Backpackers nylon tent—4-man size (*or* individual ponchos from the other lists may be snapped together and tied to make a lean-to shelter)

Basic Adult-Member List

All items on this list must be packed in the quantity shown for *each* adult family member, aged *12 or older.* Pack them in each individual's own pack or duffel bag along with his share of items from the Basic Family-Unit List.

Number On Hand	Number Needed	Item
_____	1	Large, light-weight backpack & frame *or* GI duffel bag with shoulder straps
_____	1	Light-weight nylon sleeping bag (Recommend 2½-pound Hollowfill II fill with a flannel insert, as it is cheaper than down and will not lose its loft when wet. Adapt your bags to the local climate.)
_____	1	Quart aluminum *or* GI plastic canteen with cup and cover
_____	1	Hunting knife with case (*or* pocket knife for 12 to 16-year-olds)
_____	1	Waterproof match case filled with "hurricane" matches
_____	1	Nylon GI poncho with snaps and grommets
_____	1	GI *or* small aluminum mess kit with utensils
_____	1	Metal whistle with string (for signaling)
_____	1	Pair heavy-duty leather work gloves
_____	1	Pair inexpensive polarized plastic sunglasses
_____	1	One-half roll of toilet paper in Ziploc bag
_____	1	Bar of hand soap in plastic soap case
_____	1	Old large bath towel
_____	1	Toothbrush in plastic toothbrush case
_____	1	Medium-sized tube of flouride toothpaste
_____	1	Medium-sized tube of shampoo concentrate
_____	1	Metal mirror
_____	1	Pocket comb
_____	1	Pair of old comfortable heavy-duty walking boots *or* shoes (GI combat-type or hunting boots)
_____	1	Medium weight jacket with hood *or* with ski cap
_____	1	Pair of heavy-duty jeans (Levi-type)

Basic Adult-Member List (continued)

Number On Hand	Number Needed	Item
______	1	Leather belt (optional: GI pistol belt to carry canteen, etc.)
______	1	Long-sleeve flannel shirt *or* sweater
______	3	Complete sets of underwear
______	3	Pair of heavy-duty socks
______	3	Handkerchiefs

For men who shave, add:

Number On Hand	Number Needed	Item
______	1	Lightweight razor with 10 blades
______	1	Can shaving lather (light and concentrated)
______	1	Small bottle plain after-shave lotion

For women who menstruate, add:

Number On Hand	Number Needed	Item
______	1	Small box of tampons or sanitary napkins

Basic List for Children Ages 8 to 11

All items on this list must be packed in the quantity shown for *each* child in the family, aged 8 to 11. Pack in the individual's own pack along with his share of basic family items.

Number On Hand	Number Needed	Item
_____	1	Small, lightweight backpack and frame (sized to age and strength of each child)
_____	1	Lightweight nylon sleeping bag (recommend 2½-pound Hollowfill II fill with flannel insert)
_____	1	One-quart aluminum *or* GI plastic canteen with cup and cover
_____	1	Pocket knife
_____	1	Waterproof match case filled with "Hurricane" matches
_____	1	Nylon GI poncho with snaps and grommets
_____	1	GI *or* small aluminum mess kit with utensils
_____	1	Metal whistle with string (worn around neck)
_____	1	Pair of old mittens (nylon with leather palms)
_____	1	Pair inexpensive polarized plastic sunglasses
_____	1	One-half roll of toilet paper in Ziploc bag
_____	1	Medium-sized bath towel
_____	1	Toothbrush in plastic toothbrush case
_____	1	Pocket comb
		Pair old comfortable heavy-duty walking boots *or* shoes (GI combat-type or hunting boots)
_____	1	Heavy *or* medium-weight jacket w/hood *or* cap
_____	1	Pair of heavy-duty jeans (Levi-type)
_____	1	Leather belt (optional: GI pistol belt to carry canteen)
_____	1	Long-sleeve flannel shirt *or* sweater
_____	3	Complete sets of underwear
_____	3	Pair of heavy-duty socks
_____	3	Handkerchiefs

Note: Children in this group must share the hand soap, toothpaste, shampoo, and metal mirror carried by family members in the adult group.

Basic List for Children Ages 4 to 7

All items on this list must be packed in the quantity shown for *each* child in the family, aged 4 to 7. Pack in the individual's own pack. Some children may also be able to carry some of the basic family items, including food items.

Number On Hand	Number Needed	Item
______	1	Small knapsack (sized to age and strength of each child)
______	1	One-quart aluminum *or* GI plastic canteen with cup and cover
______	1	Nylon GI poncho with snaps and grommets
______	1	Set of utensils (knife, fork and spoon)
______	1	Metal whistle with string (worn around neck)
______	1	Pair of old mittens (nylon with leather palms)
______	1	Pair inexpensive polarized plastic sunglasses
______	1	Small hand towel
______	1	Toothbrush in plastic toothbrush case
______	1	Pair of old comfortable hiking boots *or* walking shoes or heavy-duty sneakers
______	1	Heavy *or* medium weight jacket with hood *or* with cap
______	2	Pair of heavy-duty jeans (Levi-type)
______	1	Leather belt
______	2	Long-sleeve flannel shirts *or* sweaters
______	4	Complete sets of underwear
______	4	Pair of heavy-duty socks
______	1	Small old stuffed animal *or* small story book *or* coloring book and crayons

Note: Children in this group must share the sleeping bag of a family member in the adult or 8-to-11 age group. They will also use the plates and cups or bowls found in the family cook kit. They, like the 8-to-11 group, must also share the hand soap, toothpaste, shampoo, and metal mirror carried by the adult group. They do not carry their own toilet paper, pocket knife, matches or comb.

Basic List for Babies
Ages 0 to 3

All items on this list must be packed in the quantity shown (unless otherwise indicated) for *each* child in the family, aged 0 to 3. Distribute throughout all other family members' packs.

Number On Hand	Number Needed	Item
_____	1	Metal whistle with string (worn around the neck of child old enough to walk off)
_____	1	Pair warm knit mittens
_____	1	Large bath towel
_____	1	Pair old walking shoes (*or* warm bootees if not walking)
_____	1	Warm baby blanket
_____	1	Heavy-weight jacket with hood *or* with knit ski-type cap
_____	2	Pair blanket sleepers with feet (*or* bunting bag for small baby)
_____	3	Pair of old coveralls *or* long pants
_____	3	Pair long sleeved shirts *or* sweaters
_____	6	Complete sets of underwear *or* 1 dozen regular Curity diapers and 2 pair of plastic pants and 1 pair of safety pins. Also, for those still in diapers, *add* 1 small tube of diaper rash ointment and 1 small can of baby powder (per family, *not* per child). Do not forget 6 undershirts for those in diapers. Also, you may wish to add 1 small cloth-backed rubber pad for each child not yet toilet trained to help keep the sleeping bag they share dry.
_____	4	Pair of heavy-duty warm socks
_____	1	Small package of Wet-Ones-style cleaning napkins
_____	1	Small child's toy
_____	1	*Optional:* Back-pack baby carrier (Gerry-type) *or* for small babies a front-carry baby sling. If you elect to use this item, you must be prepared to *delete one pack* from your family, unless a *light* pack is used with the

Basic List for Babies Ages 0 to 3 (continued)

front sling. Be prepared to improvise a pole drag to carry your packs, if necessary, when you have a child that must be carried. You could also build a small handcart out of bike wheels. Use your ingenuity if you have to carry your children and your emergency survival packs, too.

For a child still on the bottle or nursing, add:

______	2	Plastic baby bottles with nipples and caps
______	1	Baby bottle brush (per family, *not* per child)
______	14	Cups non-fat dry milk in plastic bag (Mix 1 cup to 1 quart of water to yield 1 quart of milk per day. Supplement with other available liquids, mashed solids made from emergency food supply, and the liquid vitamins in the Basic Family-Unit List. This milk is in addition to the basic family supply and should be added even if nursing, as a mother's milk supply can be affected by the shock of crisis & diet change.)

Note: Children in this group do *not* carry a pack, even if they can walk. They do not carry their own water or eating utensils. They will share a sleeping bag with an adult or child in the 8-to-11 age group. To avoid the danger of suffocation, very small babies should *not* share a sleeping bag with an adult or large child. They should be placed in a sleeping bag with small children. In essence, they must receive all of their care and support from the older members of the group and all personal-care items must be shared.

Appendix

Suggested Survival Literature

In Time of Emergency, A citizen's Handbook on Nuclear Attack and Natural Disasters, Department of Defense, Defense Civil Preparedness Agency, Booklet #H-14, Revised, 1977. (Available through your local county Civil Defense offices. They have other useful survival/disaster literature also.)

Passport to Survival, Chapters 13 & 14 (Outdoor Survival), Esther Dickey, Bookcraft, Inc., Publishers, Salt Lake City, Utah, 1969.

How to Be Prepared, Pages 43 to 52 (Primative Survival), Roland Page, Jefferson House, Publishers, Provo, Utah, Revised 1974.

A good poisonous plant guide. One was published in 1966 by Wyeth Laboratories, a Division of American Home Products Corporation, N.Y., N.Y.

Skills for Survival—How Families Can Prepare, Esther Dickey, Horizon Publishers, Bountiful, Utah, 1978.

The ABC's of Home Food Dehydration, Barbara Densley, Horizon Publishers, Bountiful, Utah, 1975.

Home Food Dehydrating—Economical "Do-it-Yourself" Methods for Preserving, Storing and Cooking, Jay and Shirley Bills, Bountiful, Utah, 1974.

New Concepts in Dehydrated Food Cookery—Hundreds of New Ideas and Tested Recipes for Enjoying Home Dehydrated Foods, Barbara Densley, Horizon Publishers, Bountiful, Utah, 1982.

Fun With Fruit Preservation—Leather Drying, and Other Methods, Dora D. Flack, Horizon Publishers, Bountiful, Utah, Revised 1980.

The Best of Woodsmoke—A Manual of Primitive Outdoor Skills, compiled by Richard L. Jamison, Horizon Publishers & Distributors, Inc., Bountiful, Utah, 1982.

You Can Stay Alive—Wilderness Living and Emergency Survival, Larry J. Wells and Roger D. Giles, Horizon Publishers, Bountiful, Utah, 1981.

Mountainman Crafts and Skills—An Illustrated Guide to Clothing, Shelter, Equipement and Wilderness Living, David Montgomery, Horizon Publishers, Bountiful, Utah, 1980.

Cooking With Wild Plants—How to Recognize and Prepare Edible Wilderness Plants of the Rocky Mountains, Alan Briscoe, Horizon Publishers, Bountiful, Utah, 1979.

Beginners Guide to Family Preparedness, Rosalie Mason, Horizon Publishers, Bountiful, Utah, 1977.